Doomsday Scenarios: Separating Fact from Fiction™

CATASTROPHIC CLIMATE CHANGE AND GLOBAL WARMING

Frank Spalding

rosen publishing's
rosen central®
New York

For Mr. Prospero

Published in 2010 by The Rosen Publishing Group, Inc.
29 East 21st Street, New York, NY 10010

First Edition

Library of Congress Cataloging-in-Publication Data

Spalding, Frank.
Catastrophic climate change and global warming / Frank Spalding.—
1st ed.
p. cm.—(Doomsday scenarios: separating fact from fiction)
Includes bibliographical references and index.
ISBN 978-1-4358-3562-7 (library binding)
ISBN 978-1-4358-8526-4 (pbk)
ISBN 978-1-4358-8527-1 (6 pack)
1. Climatic changes. 2. Global warming. 3. Disasters. I. Title.
QC903.S63 2010
363.738'74—dc22

2009019571

Manufactured in Malaysia

CPSIA Compliance Information: Batch #TWW10YA: For Further Information contact Rosen Publishing, New York, New York at 1-800-237-9932

On the cover: This image of New York City being devastated by rising seas comes from the environmental disaster film *The Day After Tomorrow.*

CONTENTS

INTRODUCTION 4

CHAPTER 1 8
What Is Climate Change?

CHAPTER 2 21
Catastrophic Climate Change Scenarios

CHAPTER 3 33
A Rational Look at the Facts

CHAPTER 4 42
Climate Solutions

GLOSSARY 54

FOR MORE INFORMATION 55

FOR FURTHER READING 58

BIBLIOGRAPHY 60

INDEX 63

Introduction

Imagine this: Ten years from now, the president of the United States sits down to deliver a televised address to the country. Every major media outlet is covering the speech, as the president is expected to unveil an ambitious campaign to overhaul U.S. environmental policies. This plan will severely restrict the use of fossil fuels and dirty energy sources, such as coal, and force the automotive industry to create solar and electric automobile hybrids. It will also impose much stricter emissions standards on industry and create mass transit infrastructure in every American city.

The president tells the nation that the effects of global warming, while currently mild, will become far worse. Scientists all over the world have been working together for years to create a projection of what will happen to Earth's climate if we continue to pollute at our present rate. Their findings are alarming. They predict that the world will experience a total climate catastrophe. This catastrophic climate change will lead to rising sea levels, massive famines and droughts, the melting of arctic ice sheets, the transformation

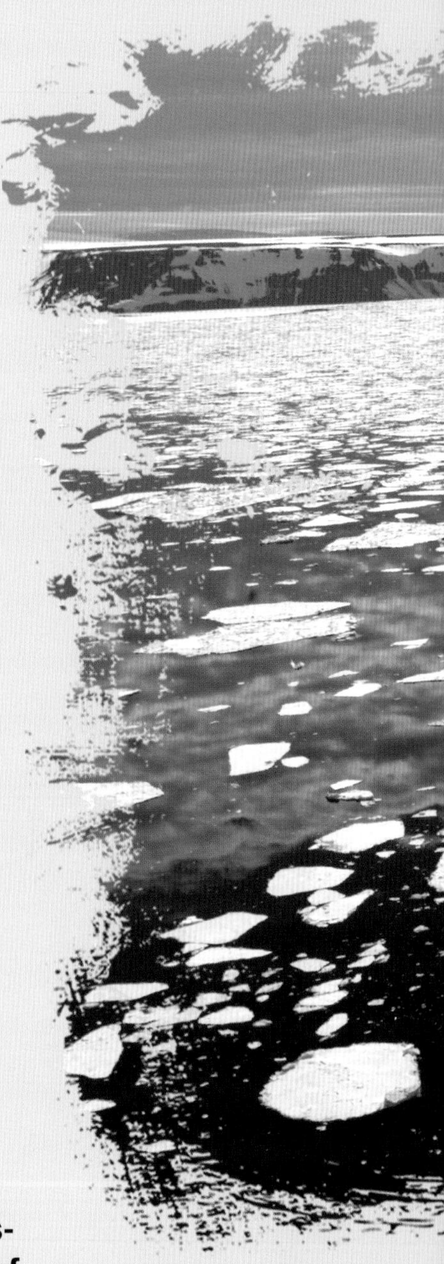

Glaciers and other ice masses in the ocean are diminishing as Earth's temperature rises. This sea ice is floating in the Arctic Ocean, off the Russian coast.

of farmland into desert, acid rain, and the extinction of numerous animal species. It will also result in increasing rates of cancer from ultraviolet (UV) radiation, and unpredictable, dangerous weather. In short, the climate scientists project that Earth could be rendered virtually uninhabitable if we continue on our present course.

The president's bill, if passed, might help prevent this grim future. It would also have profound, far-reaching effects on U.S. business and industry. The president's political opponents target the bill, saying it will harm the economy and that the science the bill is based on is faulty and unproven. The bill is voted down.

As the years pass, the effects of climate change become increasingly difficult to ignore. The rising seas threaten coastal areas, forcing low-lying cities to construct levees to protect themselves from the waters that encroach on them. Oceans become more acidic, and marine animals like fish, dolphins, and jellyfish begin to die off in large numbers. Some areas of the world are devastated when rising temperatures cause droughts and famines. Many people die. Violent conflicts erupt between the survivors, who fight over scant resources. The weather becomes unpredictable, and giant storms cause massive damage to cities and towns. Large parts of the American Southwest and Midwest become virtual deserts, dustbowls where nothing will grow.

The same panel of scientists reconvenes. Their second assessment of humanity's negative effect on Earth's climate is even harsher than the first. According to their figures, climate change is occurring faster and faster. Melting permafrost has released large amounts of the greenhouse gas methane into the air, speeding up the process of global warming. Much of

Earth's rainforests, which once helped slow the greenhouse effect by absorbing the greenhouse gas carbon dioxide, have been destroyed to make room for farmland. Global warming is spiraling out of control.

The member states of the United Nations convene and pass emergency measures to slow this warming by drastically reducing world carbon emissions. In a surprising show of solidarity, every single member state agrees to this—even large, populous, industrialized nations like the United States, Russia, China, and India. Unfortunately, it's too late. Earth's climate has been irrevocably altered. Scientists scramble to find ways to repair the damage, but the problem is now too big. If only we had acted sooner, humanity could have enjoyed a long, glorious future on a lush, green, life-nurturing planet. But we didn't. As a result, humankind's days on Earth are numbered.

Chapter 1

What Is Climate Change?

Changes to the climate affect every single person on Earth.

For all humankind's miraculous achievements, we are still an essentially fragile life-form. We can only live within a fairly narrow temperature range, and we need similarly fragile natural resources like fresh water and oxygen to survive. Earth is the only planet in the universe able to support human life.

Most scientists and experts believe that modern human beings have only existed for approximately 130,000 years. In the grand scheme of things, we are a relatively young species. Dinosaurs, by contrast, existed on Earth for more than one hundred million years before their eventual extinction. During our short time on this planet, we have invented languages, discovered agriculture, created great works of art and literature, and built inspiring monuments and cities. We have used science to discover the laws of nature, devoted ourselves to profound spiritual and religious practices, and even traveled into space. Despite the many destructive wars that have been fought between peoples, the history of humanity is that of a species that has flourished and thrived, growing and advancing with remarkable speed, expanding its reach and knowledge at an ever-increasing rate.

Modern Science, Modern Discoveries

Science has helped us both understand the natural world and examine our impact upon it. Scientists first took notice of Earth's gradual warming approximately one hundred years ago. At that time, the extent of our impact on the world was not entirely clear. The first serious research into global warming began approximately fifty years ago.

As scientists began to realize that Earth's climate was changing mainly due to human activities, people around the world became concerned. In 1988, the United Nations Environment Programme and the World Meteorological Organization formed an organization to study global warming. This organization, the Intergovernmental Panel on Climate Change (IPCC), is the most significant body of experts ever convened to examine the effects of climate change. The IPCC assesses climate data and scientific reports from hundreds of scientists around the world to produce neutral, objective reports on the state of the world's climate.

The IPCC (http://www.ipcc.ch) compiles reports on climate change based on the work of many scientists. These reports are the most comprehensive studies on climate change to date.

In 2007, the IPCC won the Nobel Peace Prize for its work. That same year, the panel released its latest findings on global climate. This report concluded that there was no question that global warming was occurring, and it was almost certainly being caused by human activity.

Progress and Industry

It is believed that humanity's impact on global climate is tied to one of our greatest achievements: the Industrial Revolution.

During the Industrial Revolution, which began during the mid-eighteenth century in England and continued through the mid-nineteenth century, new technologies changed the way that people in the western world lived and worked. Agricultural advances made the mass production of food possible for the first time. New technology also allowed people to mass-produce goods like textiles and machine parts in factories. Railroads changed how people traveled, and the telegraph revolutionized communication. These great technological advances soon spread around the world. People began moving to rapidly modernizing cities, where they sought jobs in the emerging industries created by these new technologies.

Today, industrialized production makes our way of life possible. Our stores are full of affordable toys, clothes, and household goods made in factories across the globe. Regardless of the season, supermarkets in the United States are stuffed with a dazzling variety of exotic and international fresh foods. Cars and other modes of personal transportation are abundant and affordable. Advances in science and medicine have allowed us to enjoy a much longer lifespan. We have also achieved a much higher average standard of living than at any other time in history.

In the mid-twentieth century, scientists began assessing the impact of all this industrialized human activity on the environment. Fossil fuels like coal and oil might power factories and transportation, but they also release carbon dioxide into the air. Livestock in large industrialized farms provide the world with food, but they also release a great deal of methane into the atmosphere. There are many human activities that release these and other greenhouse gases into the air. Ultimately, these emissions lead to climate change.

The Greenhouse Effect

To understand how human beings can affect the climate, we first have to understand a natural process known as the greenhouse effect. The greenhouse effect is an important natural process that existed well before human beings first appeared on the planet. In fact, it makes life on Earth possible.

When sunlight passes through the atmosphere and reaches our planet's surface, some of its energy is absorbed by the Earth. The sunlight warms Earth's surface, and this causes infrared radiation to be released. Some of this radiation exits through the atmosphere, vanishing into space. But some of it is trapped by certain greenhouse gases present in our atmosphere, such as carbon dioxide (CO_2). The energy trapped by greenhouse gases warms the atmosphere, in turn causing Earth's surface temperatures to rise.

Without the greenhouse effect, Earth would be too cold to sustain life. But if the greenhouse effect spins out of control and begins trapping too much heat, it might warm the planet too much. Consider Earth's neighbors Mars and Venus. The greenhouse effect is negligible on Mars. As a consequence, the planet is too cold for human life. On Venus, however, the greenhouse effect is very pronounced. The surface of that planet is so hot that it would cause any known life-form to evaporate instantly.

It may seem incredible that human beings have the power to alter our planet's greenhouse effect and the ancient workings of Earth's upper atmosphere. But it appears that we do. About 1 percent of Earth's atmosphere is made up of

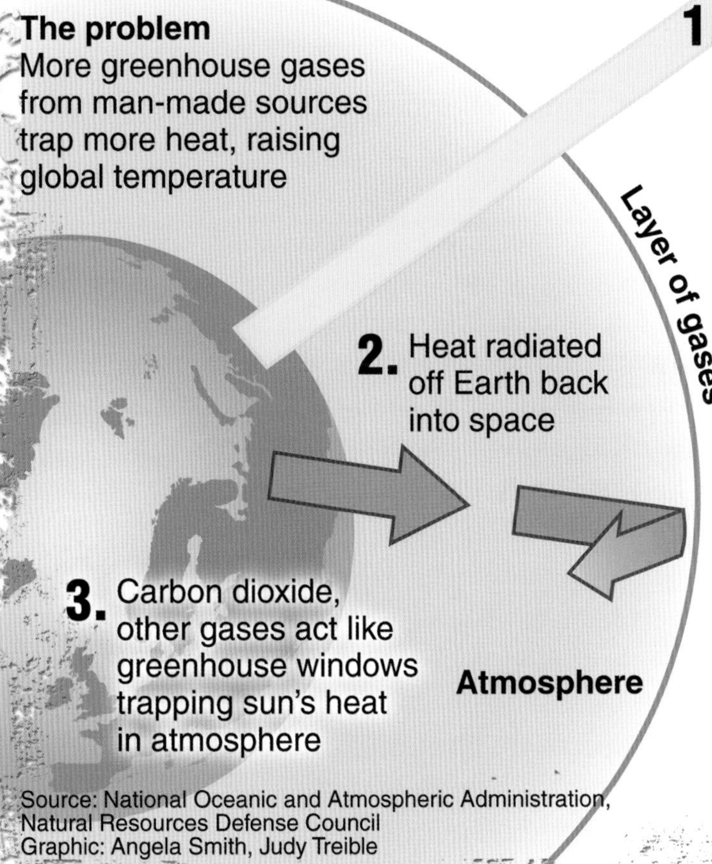

This illustration from the National Oceanic and Atmospheric Administration (NOAA) shows how global warming works.

greenhouse gases. These include carbon dioxide, methane, nitrous oxide, and water vapor, among others. The greenhouse effect currently keeps Earth's average temperature at a relatively balmy 59 degrees Fahrenheit (15 degrees Celsius). An increase in the amount of greenhouse gases in the atmosphere can cause the greenhouse effect to increase. The end result is that temperatures on Earth's surface also rise.

Greenhouse Gases

In the United States, approximately 50 percent of all consumable energy is generated by coal-burning power plants. Another 20 percent is derived from natural gas. When coal and natural gas burn, they release greenhouse gases into the atmosphere. In fact, all fossil fuels release greenhouse gases when they burn. Greenhouse gases are also released by livestock (like cows, sheep, and pigs), landfills, and melting permafrost. The United States is the world's second-largest producer of greenhouse gases, having recently been surpassed by China.

Currently, there are about 6.7 billion human beings on Earth. By the year 2040, it is estimated that there could be as many as nine billion. To put this in perspective, the

This coal-burning power plant in Conesville, Ohio, emits pollution into the air. Coal-fueled power plants have a negative impact on Earths' atmosphere. They contribute heavily to climate change.

world's human population was only about three billion in 1960. Human activity creates greenhouse gases. The more humans there are, the more greenhouse gases are released into the atmosphere.

Carbon Dioxide

Carbon dioxide is a common greenhouse gas. When you breathe in oxygen, you exhale carbon dioxide. The carbon dioxide exhaled by humans and animals is absorbed by plants, which use it to photosynthesize. During photosynthesis, plants use the energy of sunlight to convert carbon dioxide into sugars and other organic compounds. At the end of this process, plants emit oxygen, which humans and animals breathe, beginning the cycle all over again. This process is part of a phenomenon known as the carbon cycle.

The carbon cycle is one of the processes that keep Earth's atmosphere, oceans, and land balanced and healthy. Plants help remove carbon dioxide from the atmosphere, reducing the levels of the heat-trapping gas that can build up there. In addition to absorbing potentially harmful carbon dioxide, plants supply us with life-giving oxygen. When this carbon cycle is disrupted, global warming results.

Fossil fuels are composed of carbon that has been stored in the Earth for millions of years. It is believed that they're made from the compressed remains of ancient zooplankton and algae. Carbon is the most common element in the universe—in fact, it is known as the building block of all life. Fossil fuels are responsible for about 90 percent of the world's energy. When fossil fuels are burned, this carbon—which has been stored for millions of years within the planet and thus has been kept out of the atmosphere where it would trap heat—is suddenly released into the atmosphere. Human beings release an estimated 8 billion metric tons of carbon a year into the atmosphere, about 75 percent of it in the form of burnt fossil fuels. More than half of this is eventually absorbed by vegetation. The rest, however, lingers in the atmosphere, trapping heat.

Vegetation plays a crucial role in regulating Earth's climate. Unfortunately, large swaths of forest are cut down every day. Whether it's to provide wood for construction or to create fields for farming, the destruction of Earth's tree and plant life is bad news. Deforestation results in there being less vegetation to take part in the carbon cycle, meaning that more of this carbon dioxide remains in the atmosphere, trapping heat and altering the climate.

Landfills such as this one produce greenhouse gases as the organic matter in them decays.

Methane and Nitrous Oxide

The greenhouse gas methane is emitted by a number of natural sources, such as soil deposits, animal emissions (the release of gas and solid waste), swamps, and oceans. These processes account for approximately one-half the amount of methane released into the atmosphere. Human beings generate the rest. Activities like refining natural gas and other fossil fuels, processing wastewater, rice farming, and the raising of livestock all produce methane. A great deal of methane is also generated when organic matter decays. This means that swamps and wetlands release methane, as do our massive landfills (garbage dumps).

Human activities like agriculture, the burning of fossil fuels, and waste processing also produce the greenhouse gas nitrous oxide.

Other Greenhouse Gases

Ozone is a greenhouse gas as well, but it has varying effects at different levels of the atmosphere. High up, in the stratosphere, ozone is helpful. The ozone layer blocks UV light, which can cause sunburn and, over time, skin cancer. But lower down, in the troposphere, ozone is a pollutant. It is harmful to people with respiratory problems, and it traps heat at low altitudes.

Other greenhouse gases are very damaging to the ozone layer. These are high global warming potential (GWP) gases. They are not released in as great a quantity as other greenhouse gases, such as carbon dioxide. But their effect on the atmosphere can be much more serious. High GWP gases are

Earth's Temperature Cycles

"Climate change," also referred to as "global warming," is a simple term describing a very complex process. Twenty thousand years ago, Earth's climate was much cooler than it is today. Scientists refer to this period as an "ice age." During the ice age, most of North America was covered with large sheets of glacial ice.

Ice ages have occurred periodically throughout Earth's history. Some have lasted for tens of millions of years. The more temperate period between ice ages is also long—hundreds of millions of years. Earth has experienced several of these warm periods, during which conditions were much warmer than they are today. In other words, Earth's temperature fluctuates over time in a sort of cycle. The changes in Earth's temperature may be caused by a number of factors, including changes in solar radiation, the composition of atmospheric gases, ocean circulation, and other natural variables.

As scientists understand more about Earth's past climate changes, they are better able to project how our climate might change in the future. Today, scientists are mostly concerned with the impact of human activity on global climate and how it might affect the normal cycle of temperature fluctuations.

generated by many industrial processes, such as aluminum manufacturing. Although there has been a widespread effort to limit their use in the United States, GWP gas emissions have been steadily climbing.

What Is Feedback?

There are many ways that people produce emissions of global warming gases. Global warming and climate change are complicated, however. Gauging the impact of climate change is more complicated than simply measuring the degree of pollution that human beings pump into the atmosphere. The causes of climate change can create a feedback loop, which worsens the effect of global warming. What this means is that our activities can set in motion larger consequences that create a sort of snowball effect.

What is feedback? Feedback describes a unique cause-and-effect relationship. In climate science, the word "feedback" describes how a seemingly small shift in climate can cause far more dramatic climate changes. For instance, glacial ice reflects sunlight back into outer space, beyond our atmosphere. This reflected sunlight is not absorbed by Earth. This means that it doesn't generate infrared radiation that can be trapped by greenhouse gases. The rise in temperatures on Earth's surface in recent decades, however, has resulted in the reduction in size and extent of the planet's ice masses. The less ice there is to reflect sunlight, the more infrared radiation that Earth absorbs and emits back into the atmosphere, where it becomes trapped by the greenhouse gases, resulting in more climate change.

The feedback effect of climate change is particularly worrisome. Considering the degree to which our modern, industrialized way of life contributes to climate change, reversing this process will be difficult—if it's possible at all. Some scientists fear that we may have already passed a point of no return.

Chapter 2

CATASTROPHIC CLIMATE CHANGE SCENARIOS

Many large cities, like New York City and its island borough of Manhattan, are built near large bodies of water. Rising seas could one day flood these cities.

Some people have expressed happiness at the prospect of a warming planet. They mistakenly envision a planet basking in an endless summer, where the weather is perfect, winter has been banished forever, and food and water supplies remain plentiful. These people believe that global warming won't be a major problem or serious disruption to life on Earth. Yet it's clear that Earth's changing climate will not turn the entire planet into a tropical paradise.

Why not? The U.S. Environmental Protection Agency (EPA) predicts that Earth's temperature might rise by 3.2 to 7.2°F (1.8 to 4.0°C) by the year 2100. In fact, it's estimated that Earth's average temperature only increased by about 1 to 1.7°F (0.56 to 0.95°C) from 1906 to 2005. At first glance, that doesn't seem like too much of an increase. But our climate is delicate. So even small changes have far-reaching effects.

Some of these changes could initially seem positive. For instance, people living in regions of the world where it is brutally cold and infertile for much of the year will be able to enjoy warmer weather and a longer growing season. However, in the long term, far more of these climate changes are negative.

It is widely believed that weather will become more unpredictable, with an increase in dangerous, destructive, and deadly storms. Heat waves will cause people to die in the summer months. Diseases that thrive in warmer weather may spread to regions of the world where they previously could not survive. Oceans will become more acidic, and their levels will rise, swallowing miles and miles of coastal land and entire island nations. There will be famines, water shortages, and droughts. These changes will prove to be deadly for many species of animals—and possibly human beings as well.

Tornadoes can cause a great deal of damage to houses and other structures. This house in Murphysboro, Illinois, was destroyed by the Tri-state Tornado of 1925.

Extreme Weather

Global climate change might lead to stronger hurricanes and tornadoes. Severe storms can kill thousands of people and cause millions, if not billions, of dollars in property damage. Currently, the United States experiences about a thousand tornadoes a year. Places like the Netherlands, Australia, Bangladesh, New Zealand, South Africa, and the United Kingdom are also prone to tornado activity. While some tornadoes last for more than an hour, most only last ten minutes or less. But even a short-lived tornado can create catastrophic damage, destruction, and death.

The deadliest tornado in U.S. history occurred in March 1925. It was known as the Tri-State Tornado. It was so named because it passed through Missouri, Illinois, and Indiana. By the time the Tri-State Tornado had dissipated, 695 people had been killed. This is not the deadliest tornado on record, however. Tornadoes in Bangladesh have been even more deadly. A 1989 storm killed more than one thousand people there, causing enormous devastation.

Tornadoes can produce winds that generally range from 40 to 110 miles per hour (64 to 177 km/hr). If a serious one were to touch down in a major U.S. city—such as Chicago, Illinois, for instance, which is in a "tornado alley"—the damage could be staggering, causing millions of dollars in damages. Even worse, tens of thousands of people

The Netherlands (Holland) is located below sea level but remains habitable, thanks to a series of dams and dikes that keep the water at bay. Rising seas could greatly endanger this country.

could be killed. The United States has more tornadoes than any other country in the world. It will be hit particularly hard if global warming produces more numerous and powerful storms.

Melting Ice, Rising Oceans

There is a great deal of glacial ice on Earth. This ice helps regulate and moderate the planet's temperature by reflecting the sun's rays back into space. Ice also stores much of the world's fresh water. With global warming, as surface temperatures increase and the ice melts, Earth's seas will rise. This will endanger many islands and coastal communities, as well as the world's many cities that are built near the water. New York, New Orleans, Amsterdam, and many other cities could drown under the rising seas. Miles of valuable coastline could be lost. Small island nations that are only a few feet above sea level could be completely submerged. People living in these areas will have to move, creating mass migrations and strains on resources in the countries to which they flee. Businesses and industries will have to be shut down and relocate or disappear altogether.

Rising seas can contaminate freshwater sources, such as lakes and rivers, with salt. Should a tipping point be reached, the large amounts of ice contained in Antarctica and Greenland could melt, which would be disastrous. Polar bears that count on polar ice are already seeing their habitats destroyed. In fact, scientists are finding that polar bears are already drowning as they swim greater and greater distances between ice floes and sheets where they hunt and rest. They were recently declared an endangered species.

Ice isn't only found floating in the ocean. There are large areas of Earth where the ground is frozen year-round. This is known as permafrost. Trillions of metric tons of organic matter are frozen solid in Earth's permafrost. The warming globe is causing this permafrost to thaw. Permafrost stores a great amount of carbon. When it thaws, this carbon is released into

A Way of Life in Peril

One group heavily impacted by climate change is the Inuit people. Living largely within the Arctic Circle, the Inuit have existed in harmony with their environment for centuries. However, a warming climate has increasingly disrupted their way of life, which is closely tied to the snow and ice where they live. Melting permafrost is causing roads and buildings to sink and collapse, as the land they are built on is no longer completely frozen. The Inuit are worried that, should climate change continue at its current pace, their entire way of life could disappear.

the atmosphere. These vast stores of carbon, if thawed and released into the atmosphere, could mark a dramatic and sudden contribution to global warming. In East Siberia alone, it is estimated that there are 500 gigatons (1,100 trillion pounds) of frozen carbon reserves. As discussed earlier, this can begin a feedback effect—the carbon released from the melting permafrost contributes to global warming, which then causes more permafrost to melt, which releases more carbon, and the cycle perpetuates itself. If all of our permafrost thaws, the results could be disastrous.

Tropical Diseases

Because of the way that Earth is tilted on its axis, the equator is the warmest part of the planet. The poles are the

These firefighters struggle to put out a forest fire in southern France.

coldest. Certain temperatures support certain organisms better than others. Some fear that a warming globe will encourage the spread of tropical diseases, which thrive in warmer temperatures, into new regions of the globe. In a worst-case scenario, this could pose a potentially fatal danger to billions of people who have not built up natural defenses to these diseases.

Heat Waves

In 2003, a brutal heat wave descended on Europe. Temperatures were higher than they had been in years. It is believed that these high temperatures can be directly linked to the effect that human beings have on global climate. While it is impossible to pinpoint exactly how many people died as a result of the heat wave, it's believed that about thirty thousand deaths can be attributed to it. Nearly fifteen thousand people died in France alone.

Extremely hot weather is especially dangerous to the elderly. When the human body senses that it is overheating, it works overtime to maintain a safe core temperature. This can put a great strain on elderly bodies. Often, the elderly are less aware of the heat. They don't feel it as acutely as younger people do, so they may not take precautions to cool themselves down, such as seeking out fans, air conditioners, and cool drinks.

Record-high temperatures were noted in many countries in 2003. If the decades-long trend holds, it appears that temperatures will just keep rising. Will it get too hot for people to survive someday, perhaps as soon as the end of the twenty-first century? Some climate scientists think this is a distinct possibility, especially in certain parts of the world where heat, disease, famine, drought, and desertification will become extreme.

Droughts and Famines

It's possible that higher temperatures and changing weather patterns will cause widespread drought. This, in turn, could lead to massive famine. There are places in the world, such as the southwestern United States and southern Africa, which are particularly sensitive to drought conditions. Hundreds of millions of people could be affected by diminished water supplies. They could be forced to migrate to find water, which could create conflicts in the communities that they pass through or settle in.

Coupled with Earth's rapidly increasing human population, serious water shortages could be fatal for large groups of people. As land dries up and turns into desert, there may not be enough water to provide irrigation to grow food. Right now, 99 percent of all water is not fit for human consumption. The remaining 1 percent is not available to everyone on Earth—approximately 1.2 billion people do not have access to clean freshwater. According to a United Nations report, two million of these people die every year from diseases associated with polluted water. Many of those who die are children. If climate change continues as projected, this number will only increase. Human beings simply cannot live without access to clean and reliable freshwater supplies.

This coral reef near Santa Monica, California, is undergoing a process called "coral bleaching." This occurs when the organisms that live in the coral die off as a result of rising ocean temperatures.

Species Extinction

Many animals are very sensitive to changes in their ecosystems. If global warming alters those ecosystems, they may be unable to survive. Changing temperatures will almost certainly cause many species to vanish. More than a million species worldwide could be in danger of extinction.

The increase in atmospheric carbon has made the oceans more acidic. The oceans absorb a great deal of atmospheric carbon—up to 25.4 million metric tons every day. The more carbon we pump into the atmosphere, the more the oceans absorb, and the more acidic they get. Ocean acidity is harmful to many types of marine life, such as corals, plankton, and marine snails. If organisms at the bottom of the food chain, such as plankton, die out, what will happen to organisms at the top of the food chain? By 2100, the oceans could become too acidic to support a great deal of marine life. Ocean life—one of this planet's great wonders, natural resources, and food sources—could be virtually decimated, and oceans could be dull, lifeless expanses of acidic water.

China has recently surpassed the United States as the world's largest producer of greenhouse gases. This image shows heavy traffic and the resulting smog during rush hour in Beijing.

The Last Days of Humankind?

The combined effects of drought, food shortages, violent and unpredictable storms, forced migrations, and species extinction will have a major impact on life on Earth. Gradually, the environment will become less hospitable for human beings. Will Earth be able to support the many billions of people who will be born in the coming decades? Will climate change cycle out of control, accelerating rapidly until the surface of the planet is so hot that most life-forms can no longer exist? Will large groups of people die from a lack of clean water and sufficient food supplies? Will we fight wars over dwindling water resources? What will happen to the world if millions of animal species go extinct?

The frightening answer to this is that no one, not even the climate experts, knows for sure. The web of relationships that define Earth's ecosystem is extraordinarily complex. Changes in these relationships—even ones as simple as the planet warming by a few degrees over a period of decades—can have wide-ranging, yet unpredictable, results.

The vast majority of scientists believe that human beings are responsible for accelerated climate change. The clothes you wear, the hamburger you eat for dinner, and even the paper this book is printed on were created through processes that involved the production and emission of global warming gases. Nearly every single aspect of our way of life contributes to climate change. In short, every person on Earth contributes to the warming of the planet. As the human population grows larger and larger, so does our contribution to climate change. Will this end in our destruction?

Chapter 3

A Rational Look at the Facts

Climate change affects animals' ecosystems. Here, a bull makes its way through a parched landscape in the South of France.

Climate change is a complex phenomenon. There is still a lot that we don't know about our impact on Earth's climate. Sometimes, this uncertainty can be frightening. To make matters worse, different people make different predictions about how climate change will transform the planet. It can be hard to know whom we should listen to. Because there are still questions regarding how much human activity contributes to global warming, there are people and organizations that attempt to use this gap in our understanding to promote their own agendas.

Some people think that human activity doesn't contribute to global warming at all. Others think that human activity not only contributes to global warming, but that it could also end all life on Earth. Ultimately, many people have different opinions on climate change, but very few of them are actually experts in climate science. If we want to take a rational look at this phenomenon and truly understand it, our best guide is solid, scientific evidence.

There are things that we can do to lessen our impact on the atmosphere, and there are ways that we can adapt our lifestyles to a changing world. Human beings are the most intelligent life-forms on this planet. And while we are still working to fully understand the phenomenon of climate change, we have already learned enough to predict, with some certainty, what the future will hold. This knowledge can help us learn how to prepare for and adapt to our changing climate.

A Panel of Experts

The most comprehensive study of global climate change comes from the Intergovernmental Panel on Climate Change (IPCC). Formed in 1988, the IPCC is a source of objective and nonpartisan research into climate change. The organization assesses climate data from hundreds of scientists around the world. After processing this data, it produces reports summarizing the findings. The most recent IPCC report was released in 2007, and it is the most definitive report on climate change ever produced. By drawing from such a wide variety of sources, the IPCC has produced some of the most thorough, accurate, and compelling research into climate change.

According to the IPCC's "Climate Change 2007: Synthesis Report," there is no reason to think that human beings are in any immediate danger of dying out due to climate change. It's important to understand that global warming is a slow process, taking place over hundreds, even thousands, of years.

Global Effects

With almost complete certainty, it is now known that average global temperatures will increase throughout the twenty-first century and beyond. We also know that human beings are contributing to this temperature increase.

According to the IPCC, there have been fewer cold days and more heat waves over the last fifty years than at any other time since temperatures have been recorded. Furthermore, the temperatures of the last fifty years were, on average, the highest in the Northern Hemisphere since at least 1000 CE. The IPCC believes that this warming can be directly traced back to human activity. Moreover, the last decade was the warmest one ever recorded. It can only be assumed that the record-breaking temperatures being recorded now will be surpassed in the near future.

Snow and ice cover around the world have decreased. Many ancient glaciers will not survive the twenty-first century. High temperatures have caused sea ice in the Arctic to melt at a rate of approximately 2.7 percent per decade since 1978. This melting ice has caused, and will continue to cause, sea levels to rise. In Antarctica, large chunks of ice have broken free of the main Antarctic ice mass. For instance, in 2002, a large Antarctic ice mass known as the Larsen B broke free and collapsed into the ocean. Ice shelves melt quicker in the open ocean than they do when attached to the main ice shelf.

If Earth's sea ice melts, it would have a catastrophic effect on the planet. Here, we see the Larsen B ice shelf collapsing due to rising temperatures.

Greenland, a nation largely covered by ice, is also experiencing melting. Estimates as to how much ice will be lost vary, but the process is definitely well under way.

Rising oceans will pose a threat to coastal cities and communities near the shore. However, cities and communities can adapt to rising waters. For instance, the cities of New Orleans and Amsterdam are already below sea level. They use a series of dikes, canals, levees, and drainage systems to control the water. If global sea levels rise the way they are projected to, cities like New York may eventually be below sea level, too. Just as Amsterdam and New Orleans have managed to survive (sometimes just barely) while being surrounded by potentially flooding waters, so, too, will many other cities affected by rising oceans. While there may not be a way to save every place threatened by rising oceans, humanity as a whole will not be endangered by this. Instead, this is one of the many regrettable side effects of global warming that we will have to adapt to over the years.

We know that the oceans are becoming acidic, but we don't yet know the extent to which this will affect marine life. While very acidic oceans would be fatal to many ocean species, the oceans are acidifying very gradually. Since this process is the result of the oceans absorbing atmospheric carbon, a reduction in the amount of carbon we put into the air might be able to slow, stop, or even reverse the trend. Some species may even have time to adapt to higher levels of acidification.

Effects by Region

Climate change will have different impacts on different regions of the world. In the United States, the effects might not be disastrous. Although the United States is tornado-prone, there is no definite research that directly links an increased likelihood of tornadoes to climate change. Flooding is an issue that coastal communities will eventually have to contend with. Higher temperatures and heat waves will particularly affect cities, which are already several degrees warmer on average than suburban or rural areas. This will be true for all of North America.

These houses in Fargo, North Dakota, are being swallowed up by the flooding Red River. The effects of increasingly extreme weather and rising water levels have been felt by communities all around the world.

In South America, the impact of climate change on the Amazon rain forest and other tropical areas could be significant. The rain forest is a vital component of the carbon cycle. The lush vegetation, home to millions of animal and plant species, also absorbs a great deal of atmospheric carbon each and every day. Many of the animals living in the rain forest are sensitive to environmental changes, and there is a strong possibility that many could become extinct in the coming decades. If drought were to cause the Amazon rain forest to die off, it could release into the atmosphere the 90 billion tons of carbon it currently stores.

In Africa, water stress will be far more pronounced. According to the IPCC, 75 to 250 million people will be affected by a scarcity of water resources in the next decade or so. Changing weather could make things difficult for nonirrigated agriculture, which is agriculture that relies on direct rainfall, rather than pumped-in water supplies from aquifers or reservoirs. Small or failed harvests would place enormous stress on food supplies. Increased land aridity (dryness) will also negatively impact agriculture and food production.

Much of Asia will be heavily impacted by climate change, especially when it comes to drinking water. The IPCC estimates that freshwater will become increasingly scarce by 2050. Asia's large population—approximately four billion people—means that water stress will have a serious impact on the continent. Flooding and the spread of disease, especially water-borne disease, will also pose a threat to Asia's population.

Australia, New Zealand, and the Pacific Islands will almost certainly experience wildlife extinction, especially among offshore coral reefs. Droughts and increased aridity could impact agriculture, as well as lead to an increase in forest fires. Some offshore islands in the Pacific could have their landmasses reduced by the rising sea level, or they could disappear entirely.

Climate Change in the Developing World

In the new millennium, emerging nations like China and India have risen as impressive world powers with thriving economies. China is exempt from the Kyoto treaty because it is a developing country. Former U.S. president George W. Bush cited this exemption as the reason why the United States would not ratify (sign) the treaty. As a developing nation with an enormous manufacturing industry, China has prioritized economic expansion over regulating emissions and protecting the environment. It's primarily a coal-powered country, with power plants and factories burning tons and tons of coal every day. It is estimated that, should China continue at its current rate, it will burn more coal than all other nations combined within the next twenty years. This could have an enormous impact on world climate change.

India, another developing nation that consumes a great deal of coal, also poses a threat to Earth's climate. Currently, India is the second most populous nation in the world, and its population is estimated to overtake China's in the coming decades. Combined with emissions produced by developed countries like the United States, the emissions from large developing nations could result in a climate catastrophe.

Adapting to a Changing Climate

It is believed that one of the reasons that heat waves caused so many deaths in Europe in 2003 is that many Europeans weren't used to such high temperatures and, therefore, weren't knowledgeable about ways to cope. With the proper preparations, far fewer people would succumb to the heat. As climate change continues, we will have to find ways to adapt to our new world.

According to the IPCC, we can help offset the effects of climate change in a variety of ways. For instance, water stress could be lessened by better management of our freshwater supplies (including lakes, rivers, and streams), wastewater reprocessing, and the use of desalinization technology to make saltwater drinkable. We can avoid aridity and desertification by planting more trees. More efficient farming techniques and crop irrigation can reduce food shortages. Public health initiatives can help avoid the spread of disease.

Climate change is frightening, but there is no evidence that humanity as a whole is in immediate peril. This does not mean that we can ignore climate change—nothing is further from the truth. It does mean, however, that we will find ways to cope with the changes occurring in our climate. If we begin now, it's possible that we will have a bright future ahead of us.

Chapter 4

CLIMATE SOLUTIONS

People all around the world are concerned about Earth's climate. Here, citizens of Auckland, New Zealand, participate in a Global Day of Action on climate change.

The human mind is a remarkable thing. Our species' capacity for inventiveness knows no bounds. Although our progress has contributed to climate change, there is no reason why our inventiveness can't reverse this trend. From large governmental structures to individual citizens, we can all take an active role in reversing climate change. It may not be easy at first, but we can do anything if we put our minds to it.

Global Solutions

We have already taken an important step in fighting climate change by identifying the problem and its causes. Every year, we learn more about what causes climate change, and every year we come up with more possible solutions. Unfortunately, as good as we are at identifying the problem and devising ways to solve it, we haven't yet implemented a concrete, comprehensive plan that would fix it.

Realistically, any solution to climate change would have to be conducted on a global scale, with the full participation of all the world's nations and peoples. In 1997, the international community worked together to create a piece of legislation that would coordinate countries worldwide in the fight against climate change. This legislation, known as the Kyoto Protocol, went into effect in 2005. While the United States did not sign the treaty, the majority of the world's countries have.

The changes that the Kyoto Protocol calls for seem relatively minor. But some countries feel that even these small reductions in greenhouse emissions will negatively impact their economies. The treaty calls for its signatories to reduce their greenhouse emissions to a level that is 5 percent below what they were emitting in 1990. If all of the world's largest polluters were to ratify the treaty, it would not only be a powerful tool in reducing climate change, but it would also be a strong signal that the world is ready to come together, share the burden, and make further sacrifices for Earth's future.

Some people feel that the Kyoto Protocol, which is due to expire in 2012, is too limited in scope to solve the world's

Working to stop climate change needs to be an international effort. Here, delegates from a number of countries attend the UN Climate Change Talks in Bangkok, Thailand.

problems. A lot of time has passed since its inception in 1997, and many feel that a new global climate policy would better combat climate change.

In 2007, the 190 UN member states pledged to create a new global climate treaty. The treaty is projected to be more aggressive than the Kyoto Protocol. In 2009, the United States, under newly elected president Barack Obama, signaled that the nation would now be willing to cooperate with the United Nations on any new treaty that emerges from these discussions. This is a powerful commitment from

Eat Locally

Although many people may not think about it, the food that they buy in the grocery store is often not grown locally. We've become accustomed to getting any kind of food at any time of the year, whether it is in season or not. Demand for certain foods means that they are shipped in, often from very far away. Transporting food long distances creates carbon emissions. You can avoid this by purchasing locally grown produce and other food products. Many communities have farmer's markets or farm stands where this food can be purchased. Buying this food not only benefits local farmers, but it also helps you do your part to combat climate change by reducing greenhouse emissions.

the world's second largest producer of greenhouse gases to do its part to protect our increasingly fragile planet and the life that it supports and sustains.

Clean Energy

Solar power is one of the most promising sources of green energy. It uses the sun's rays to generate electricity, making it a completely clean form of energy. Installing solar panels on the roof of a house can greatly reduce a homeowner's electric bill. Large solar collectors may someday power entire communities. This technology is still evolving, and the solar collectors of tomorrow could be much more efficient than the ones we have today.

Solar power is a source of clean energy that is catching on around the world. This house has solar panels on its roof that generate the required electricity.

Wind power is another source of clean energy. Although wind turbines can only function properly in parts of the country that enjoy steady and reliable winds, some states get up to 20 percent of their energy from wind power.

While this technology may never be able to provide clean energy to the entire world, it can greatly contribute to the reduction of greenhouse gases in a large number of localities and regions.

Hydroelectric power uses the power of rushing water to generate electricity. It is a clean and renewable energy source that provides the United States with about 7 percent of its total electricity. Like wind power, hydroelectric power on its own will probably never be a global solution to climate change. Not every waterway can be harnessed to generate hydropower, and not every region has access to ample, powerful, and moving water supplies. While hydroelectricity is a clean energy source, it can still cause ecological damage. Hydroelectric dams, for example, can disturb the habitat of fish, bird, and animal species in the surrounding area. This is especially true in those areas that have been deprived of their usual water supplies following the dams' construction.

Nuclear Power

Often considered to be a very dangerous source of energy, nuclear power is nevertheless put forth as a far cleaner, more abundant, and renewable source of energy than fossil fuels. France is perhaps the world's biggest proponent of nuclear power—it gets approximately 80 percent of its energy from

Some politicians support nuclear power as a solution to greenhouse gas–producing forms of energy. However, the radioactivity associated with nuclear power remains a potential danger to the surrounding environment.

nuclear power plants. Only 20 percent of the United States' energy comes from nuclear power plants.

While nuclear power doesn't generate greenhouse gases, it does create extraordinarily hazardous radioactive waste.

Storing this waste can be a problem, as it poses a severe threat to anyone exposed to it, and it can leach into soil and groundwater. Nuclear power can also be very dangerous if something goes wrong at the power plant.

For instance, in 1986, the Chernobyl nuclear power plant in Ukraine experienced what is known as a nuclear meltdown. An error caused one of the reactors to explode, releasing radioactive material into the atmosphere. Much of it contaminated neighboring countries. To date, this has been the worst nuclear accident that the world has known. Fifty-six people died in the explosion itself, and as many as four thousand people may have eventually died of radiation sickness and radiation-related cancers. The area around the Chernobyl plant is still very radioactive.

With the proper oversight, nuclear power is considered to be perfectly safe. Unfortunately, there is no way to guarantee that nuclear power plants will not experience calamities on par with the Chernobyl disaster.

Emissions Standards

In 2006, approximately one-third of global warming gases produced in the United States were caused by transport, according to the U.S. Environmental Protection Agency (EPA).

This hybrid automobile produces fewer harmful emissions than the average car. Technology and new innovations like this may help us avoid climate catastrophe.

Stricter vehicle emissions standards could reduce this. Although the technology exists to create more fuel-efficient vehicles like high-mileage, hybrid, and electric cars, most automobiles are nowhere near as energy efficient as they could be. Alternate automobile fuels, such as ethanol, are also being explored as options that might produce less harmful emissions than petroleum products.

Taking Charge of Your Future

Climate change affects us all, and we can all join the effort to stop it. We don't have to wait for global climate agreements to be signed into law by international governments before we start doing our part to fight climate change. Approximately one-third of U.S. carbon emissions arise from transportation. Using public transportation or riding a bike to school and work can cut down on your carbon emissions. You can also purchase locally grown food or locally produced goods, such as clothing or other consumer items. Much of what we buy has traveled great distances to reach us, creating greenhouse gas pollution in the process.

Many products that we buy every day are designed to be discarded. Newspapers, magazines, cardboard packaging, aluminum cans, and glass and plastic bottles are some things that can easily be recycled, rather than simply thrown away and dumped in a landfill. Recycling these products prevents waste, saves energy, and cuts down on the amount of greenhouse gas that would otherwise be used to manufacture new versions of these products from scratch.

By making the effort to conserve electricity at home, you can make a dent in your carbon emissions. The electricity that we use when turning on lights and appliances comes to

Buying locally grown produce helps local family farms stay in business. It also helps the environment by drastically reducing greenhouse gas emissions associated with the transporting of produce from distant growers to local supermarkets.

us from power plants, which generate greenhouse gases. Remembering to turn off the lights when you leave a room and turn off and unplug electronic appliances when you go to sleep can dramatically reduce your carbon footprint.

Getting Involved

If you want to take a more active role in preventing climate change, there's no reason not to start now. Plant a garden, convince your parents to buy highly efficient compact fluorescent light bulbs, recycle regularly, and learn how to effectively winterize your home to save energy. You can see what kind of organizations or clubs your school or community has that are devoted to environmental causes. Learn about environmental issues in your community. If you're interested in creating change, you can pursue academic subjects and majors that are dedicated to climate science or political science. Ultimately, you will help to create the world that you will inhabit as an adult and hand over to your children and grandchildren.

Glossary

acidification The process by which something becomes more acidic. In the case of Earth's oceans, an intake of carbon dioxide causes the water to become more acidic.

aridity A lack of moisture.

atmosphere The cloud of gases surrounding Earth.

drought A prolonged shortage of water.

emission Something released as a result of certain processes. For instance, cars emit carbon dioxide as exhaust, and the uranium used in nuclear power plants emits radiation as it decays.

extinction The death of every single organism of a certain species of plant or animal.

famine A mass food shortage resulting in widespread hunger.

fluctuations Variations in something that can be measured, such as temperature.

fossil fuels Nonrenewable energy sources (including coal, methane, and petroleum) that contain a high percentage of carbon or hydrocarbon and are formed by the decomposition of buried dead organisms over hundreds of millions of years. When burned to generate energy, they emit carbon dioxide into the atmosphere.

levee A dam used to keep an area from flooding.

livestock Cattle, pigs, sheep, and other domestic farm animals that are raised to be consumed as food or sold for profit.

marine Having to do with the water. Marine animals are those that spend their lives in oceans, lakes, or rivers.

migration The mass movement of groups of people or animals from one location to another.

solidarity Unity within a group whose members share a common cause and fellow feeling.

For More Information

Environmental Defence Canada
317 Adelaide Street, West Suite 705
Toronto, ON M5V 1P9
Canada
(416) 323-9521
Web site: http://www.environmentaldefence.ca
This Canadian organization promotes environmental protection.

Greenpeace
702 H Street NW
Washington, DC 20001
(202) 462-1177
Web site: http://www.greenpeace.org/usa
This global organization has campaigned for environmental issues since 1971.

Nature Conservancy
4245 North Fairfax Drive, Suite 100
Arlington, VA 22203-1606
(703) 841-5300
Web site: http://www.nature.org
This conservation organization identifies principal threats to marine life, freshwater ecosystems, forests, and protected areas, and then uses a scientific approach to save them.

Pew Center on Global Climate Change
2101 Wilson Boulevard, Suite 550
Arlington, VA 22201
(703) 516-4146

Web site: http://www.pewclimate.org
The Pew Center on Global Climate Change works to develop solutions to climate change.

Sierra Club
85 Second Street, 2nd Floor
San Francisco, CA 94105
(415) 977-5500
Web site: http://www.sierraclub.org
Founded in 1892, the Sierra Club is the United States' oldest environmental organization.

Union of Concerned Scientists
2 Brattle Square
Cambridge, MA 02238-9105
(617) 547 5552
Web site: http://www.ucsusa.org
The Union of Concerned Scientists works to educate the public about issues relating to climate change.

United Nations
First Avenue at 46th Street
New York, NY 10017
(212) 963-4475
Web site: http://www.un.org
An international organization comprised of nearly every sovereign nation on Earth, the United Nations uses diplomacy to find peaceful solutions to world problems.

U.S. Environmental Protection Agency (EPA)
Ariel Rios Building

1200 Pennsylvania Avenue NW
Washington, DC 20460
(800) 438-2474
Web site: http://www.epa.gov
Founded in 1970, the EPA works to shape U.S. environmental policy.

Web Sites

Due to the changing nature of Internet links, Rosen Publishing has developed an online list of Web sites related to the subject of this book. This site is updated regularly. Please use this link to access this list:

http://www.rosenlinks.com/doom/clim

For Further Reading

Brezina, Corona. *Climate Change*. New York, NY: Rosen Publishing, 2007.

David, Laurie, and Cambria Gordon. *The Down-to-Earth Guide to Global Warming*. New York, NY: Orchard Books, 2007.

Dinwiddie, Robert, and Louise Thomas. *Ocean*. New York, NY: DK Press, 2006.

Friend, Robyn, Judith Love Cohen, and David A. Katz. *A Clean Sky: The Global Warming Story*. Marina del Rey, CA: Cascade Pass, 2007.

Gershon, David. *Low Carbon Diet*. Woodstock, NY: Empowerment Institute, 2006.

Goodall, Chris. *How to Live a Low-Carbon Life*. London, England: Earthscan Publications, Ltd., 2007.

Gore, Al. *An Inconvenient Truth: The Crisis of Global Warming*. New York, NY: Penguin, 2006.

Grist magazine, and Brangien Davis, ed. *Wake Up and Smell the Planet*. Seattle, WA: Mountaineers Books, 2007.

Johnson, Kirk. *Gas Trees and Car Turds: Kids' Guide to the Roots of Climate Change*. Golden, CO: Fulcrum Publishing, 2007.

Langholz, Jeffrey, and Kelly Turner. *You Can Prevent Global Warming (and Save Money!): 51 Easy Ways*. Kansas City, MO: Andrews McMeel Publishing, 2003.

Nikel-Zueger, Manuel. *Critical Thinking About Environmental Issues: Energy*. Farmington Hills, MI: Greenhaven Press, 2003.

Silverstein, Alan. *Global Warming*. Kirkland, WA: 21st Century, 2003.

Spence, Christopher. *Global Warming: Personal Solutions for a Healthy Planet*. New York, NY: Palgrave MacMillan, 2005.

Tanaka, Shelley. *Climate Change*. Toronto, ON, Canada: Groundwood Books, 2006.

Thornhill, Jan. *This Is My Planet: The Kids' Guide to Global Warming*. Toronto, ON, Canada: Maple Tree Press, 2007.

Woodward, John. *Climate Change* (DK Eyewitness Books). New York, NY: DK Children, 2008.

Young, Mitchell. *Garbage and Recycling*. Farmington Hills, MI: Greenhaven Press, 2007.

Bibliography

Aberystwyth University. "Antarctic Ice Shelf Collapse Blamed on More Than Climate Change." *ScienceDaily*, February 11, 2008. Retrieved April 2009 (http://www.sciencedaily.com/releases/2008/02/080210100441.html).

Adam, David. "Surge in Carbon Levels Shows Vegetation Struggling to Cope." Guardian.co.uk, May 11, 2007. Retrieved April 2009 (http://www.guardian.co.uk/environment/2007/may/11/highereducation.climatechange).

American Geophysical Union. "Global Warming Could Release Trillions of Pounds of Carbon Annually from East Siberia's Vast Frozen Soils." *ScienceDaily*, June 12, 2008. Retrieved April 2009 (http://www.sciencedaily.com/releases/2008/06/080611154839.htm).

American Institute of Biological Sciences. "Thawing Permafrost Likely to Boost Global Warming, New Assessment Concludes." *ScienceDaily*, September 2, 2008. Retrieved April 2009 (http://www.sciencedaily.com/releases/2008/09/080901084854.htm).

Associated Press. "Scientists Warn of Water Shortages and Disease Linked to Global Warming." *New York Times*, March 12, 2007. Retrieved April 2009 (http://www.nytimes.com/2007/03/12/science/earth/12climate.html).

Bernstein, Lenny, et al. "Climate Change 2007: Synthesis Report. Summary for Policymakers." Intergovernmental Panel on Climate Change, November 17, 2007. Retrieved April 2009 (http://www.ipcc.ch/pdf/assessment-report/ar4/syr/ar4_syr_spm.pdf).

Bradsher, Keith. "China to Pass U.S. in 2009 in Emissions." *New York Times*, November 7, 2006. Retrieved April 2008 (http://www.nytimes.com/2006/11/07/business/worldbusiness/07pollute.html).

Bradsher, Keith, and David Barboza. "Pollution from Chinese Coal Casts a Global Shadow." *New York Times*, June 11, 2006. Retrieved April 2009 (http://www.nytimes.com/2006/06/11/business/worldbusiness/11chinacoal.html).

Brown, Paul. "Global Warming Is Killing Us, Too, Say Inuit." Guardian.co.uk, December 11, 2003. Retrieved April 2009 (http://www.guardian.co.uk/environment/2003/dec/11/weather.climatechange).

Clover, Charles. "IPCC: Lawson Wrong About Climate Change." Telegraph.co.uk, November 10, 2008. Retrieved April 2009 (http://www.telegraph.co.uk/earth/earthcomment/charlesclover/3339514/IPCC-Lawson-wrong-about-climate-change.html).

Dessler, Andrew E., and Edward A. Parson. *The Science and Politics of Global Climate Change: A Guide to the Debate*. New York, NY: Cambridge University Press, 2006.

Dow, Kristin, and Thomas Downing. *The Atlas of Climate Change: Mapping the World's Greatest Challenge*. Berkeley, CA: University of California Press, 2007.

Eilperin, Juliet. "Growing Acidity of Oceans May Kill Corals." *Washington Post*, July 5, 2006. Retrieved April 2009 (http://www.washingtonpost.com/wp-dyn/content/article/2006/07/04/AR2006070400772.html).

Emanuel, Kerry. *What We Know About Climate Change*. Cambridge, MA: MIT Press, 2007.

Henson, Robert. *The Rough Guide to Climate Change*. 2nd ed. New York, NY: Rough Guides, 2008.

Houghton, John. *Global Warming: The Complete Briefing*. New York, NY: Cambridge University Press, 2004.

Kolbert, Elizabeth. *Field Notes from a Catastrophe: Man, Nature, and Climate Change*. New York, NY: Bloomsbury USA, 2006.

Lean in Manaus, Geoffrey, and Fred Pearce. "Amazon Rainforest 'Could Become a Desert.'" Independent.co.uk, July 23, 2006. Retrieved April 2009 (http://www.independent.co.uk/environment/amazon-rainforest-could-become-a-desert-408977.html).

Maasch, Kirk A. "Cracking the Big Chill." Nova Online. Retrieved April 2009 (http://www.pbs.org/wgbh/nova/ice/chill.html).

McKie, Robin. "Scientists to Issue Stark Warning Over Dramatic New Sea Level Figures." Guardian.co.uk, March 8, 2009. Retrieved April 2009 (http://www.guardian.co.uk/science/2009/mar/08/climate-change-flooding).

Mooney, Chris. "Is Climate Change Causing an Upsurge in U.S. Tornadoes?" *New Scientist*, July 30, 2008. Retrieved April 2009 (http://www.newscientist.com/article/mg19926671.800-is-climate-change-causing-an-upsurge-in-us-tornadoes.html).

Quinn, Ben. "China's Carbon Dioxide Production Soars." Telegraph.co.uk, June 20, 2007. Retrieved April 2009 (http://www.telegraph.co.uk/earth/earthnews/3298031/Chinas-carbon-dioxide-production-soars.html).

U.S. Environmental Protection Agency. "Greenhouse Gas Emissions from the U.S. Transportation Sector 1990–2003." Office of Transportation and Air Quality, March 2006. Retrieved April 2009 (http://www.epa.gov/otaq/climate/420r06003.pdf).

Weart, Spencer R. *The Discovery of Global Warming*. Cambridge, MA: Harvard University Press, 2008.

Index

A

acidification, 6, 22, 31, 37
Amazon rain forest, 39

C

carbon cycle, 16
carbon dioxide, 7, 11, 12, 14, 16–17, 18, 26–27, 31, 37, 39, 45, 51
Chernobyl, 49
clean energy, 45–47
climate change, and developing nations, 40

D

drought, 4, 22, 29–30, 31, 39

E

eating locally, 45, 51
emissions standards and regulation, 4, 7, 40, 43, 49, 51
ethanol, 51
extinction, 6, 30–31, 32, 39

F

famine, 4, 6, 22, 29–30
feedback, 20, 27
flooding, 22, 37, 38, 39
fossil fuels, 4, 11, 14, 16, 18, 47

G

greenhouse effect, 12, 14
greenhouse gases, 6, 7, 11, 12, 14–19, 20, 43, 45, 47, 48, 51, 53

H

heat waves, 22, 28–29, 37, 41
high global warming potential (GWP) gases, 18–19
hydroelectric power, 47

I

ice melting, as a result of global warming, 4, 20, 26–27, 35–36
Industrial Revolution, 10–11
Intergovernmental Panel on Climate Change (IPCC), 9–10, 34–35, 39, 41
Inuit people, 27

K

Kyoto Protocol, 40, 43–44

L

Larsen B ice shelf, 35

M

methane, 6, 11, 14, 18

N

nitrous oxide, 14, 18
nonirrigated agriculture, 39
nuclear power, 47–49

P

permafrost, 26–27
photosynthesis, 16
polar bears, 26

R

recycling, 51, 53

S

sea levels, rising, 4, 6, 22, 26, 35–37, 40
solar power, 4, 45

T

temperature cycles, 19
tornadoes, 23–25, 37
tropical diseases, 27–28

W

wind power, 46–47

About the Author

Frank Spalding is a writer based in New York. He has written a number of titles for Rosen Publishing on urgent political and environmental issues. Spalding is a frequent traveler with a passion for global politics. He is currently embarking on a research sabbatical.

Photo Credits

Cover, p.1 © 20th Century Fox Film Corp./Everett Collection; pp. 4–5 © Sue Flood/Getty Images; pp. 8, 36 © NASA; p. 13 © Newscom; pp. 14–15 © Peter Essick/Aurora/Getty Images; p. 17 © Melissa Farlow/Getty Images; p. 21 © John Blackford/American Artists Representatives, Inc.; pp. 22–23 ©Topical Press Agency/Hulton Archive/Getty Images; pp. 24–25 © John Elk III/Getty Images; p. 28 © Jacques Munch/AFP/Getty Images; p. 30 © AP Photos; p. 31 © Christian Kober/Getty Images; p. 33 © AFP Photo/Dominique Paget/Getty Images; p. 38 © Scott Olson/Getty Images; p. 42 © Dean Treml/Getty Images; p. 44 © Saeed Khan/AFP/Getty Images; pp. 46–47 © Bernhard Lang/Getty Images; pp. 48–49 © Fred Dufour/AFP/Getty Images; p. 50 © Mark Renders/Getty Images; p. 52 © Jeff Greenberg/Getty Images.

Designer: Sam Zavieh; Photo Researcher: Marty Levick

DATE DUE

Return Material Promptly